OFFICER CORP[illegible]

TOWARDS A U.S. [illegible]
STRATEGY [illegible]
EMPLOY[illegible]

Casey Wardynski
David S. Lyle
Michael J. Colarusso

May 2010

*****

*****

Comments pertaining to this report are invited and should be forwarded to: Director, Strategic Studies Institute, U.S. Army War College, 122 Forbes Ave, Carlisle, PA 17013-5244.

*****

All Strategic Studies Institute (SSI) publications may be downloaded free of charge from the SSI website. Hard copies of this report may also be obtained free of charge by placing an order on the SSI website. SSI's website address is: *www.StrategicStudiesInstitute.army.mil.*

*****

The Strategic Studies Institute publishes a monthly e-mail newsletter to update the national security community on the research of our analysts, recent and forthcoming publications, and upcoming conferences sponsored by the Institute. Each newsletter also provides a strategic commentary by one of our research analysts. If you are interested in receiving this newsletter, please subscribe on the SSI website at *www.StrategicStudiesInstitute.army.mil/newsletter/.*

*****

This is the final monograph in a series of six monographs that analyze the development of an Officer Corps strategy. Previous volumes are:

1 - *Towards A U.S. Army Officer Corps Strategy for Success: A Proposed Human Capital Model Focused Upon Talent,* by Colonel Casey Wardynski, Major David S. Lyle, and Lieutenant Colonel (Ret.) Michael J. Colarusso, April 2009.

2 - *Talent: Implications for a U.S. Army Officer Corps Strategy,* by Colonel Casey Wardynski, Major David S. Lyle, and Lieutenant Colonel (Ret.) Michael J. Colarusso, November 2009.

3 - *Towards a U.S. Army Officer Corps Strategy for Success: Retaining Talent,* by Colonel Casey Wardynski, Major David S. Lyle, and Lieutenant Colonel (Ret.) Michael J. Colarusso, January 2010.

4 - *Accessing Talent: The Foundation Of A U.S. Army Officer Corps Strategy,* by Colonel Casey Wardynski, Major David S. Lyle, and Lieutenant Colonel (Ret.) Michael J. Colarusso, February 2010.

5 - *Towards A U.S. Army Officer Corps Strategy for Success: Developing Talent,* by Colonel Casey Wardynski, Major David S. Lyle, and Lieutenant Colonel (Ret.) Michael J. Colarusso, March 2010.

ISBN 1-58487-445-7

FOREWORD

Towards a U.S. Army Officer Corps Strategy for Success: Employing Talent is the last of six monographs focused upon officer talent management in the U.S. Army. Here, Colonel Casey Wardynski, Major David Lyle, and Lieutenant Colonel (Ret.) Michael J. Colarusso argue that the Army's current officer employment paradigm is unequal to the needs of a professional, volunteer Army facing the twin challenges of a competitive labor market and an increasingly complex global operating environment. The authors then explain the ways in which optimal employment theories, information age tools, and well-regulated market mechanisms can generate better talent matches, making the Officer Corps far more productive.

As the employment of talented officers is a necessary component of any future Officer Corps strategy, the theories discussed in this monograph merit close attention.

DOUGLAS C. LOVELACE, JR.
Director
Strategic Studies Institute

ABOUT THE AUTHORS

CASEY WARDYNSKI is Director of the Office of Economic and Manpower Analysis and an Associate Professor of Economics at the United States Military Academy, West Point, NY. In addition to creating the concepts for the "America's Army" game and the Army's pre-commissioning retention incentives, Colonel Wardynski has published in the area of military compensation policy and manpower. Colonel Wardynski earned a B.S. at West Point, a Masters in Public Policy at Harvard, and a Ph.D. in Policy Analysis from the Rand Graduate School.

DAVID S. LYLE is an Associate Professor of Economics and Deputy Director of the Office of Economic and Manpower Analysis at the United States Military Academy, West Point, NY. He has publications in the Journal of Political Economy, the Journal of Labor Economics, the Review of Economics and Statistics, the American Economic Journal: Applied, and the Economics of Education Review. Major Lyle earned a B.S. at West Point and a Ph.D. in Economics from the Massachusetts Institute of Technology.

MICHAEL J. COLARUSSO is a research analyst in the Office of Economic and Manpower Analysis at the United States Military Academy, West Point, NY. He is a retired U.S. Army lieutenant colonel and has served in a variety of military positions, to include as an Assistant Professor of History at West Point. Mr. Colarusso earned a B.A. in History from Saint John's University and a M.A. in History from the Pennsylvania State University.

SUMMARY

In the Information Age, jobs are becoming more complex, requiring employees who are agile, inventive, and empathetic. Work is increasingly characterized by high levels of task interdependence, skill specificity, and uncertainty. In addition, today's enormously competitive labor market gives educated professionals the option of seeking new employment whenever a company fails to give them sufficient voice in their work. In short, the industrial era, during which "bosses" unilaterally made employment decisions, is over.

Today, the most successful enterprises unleash the talents of their workers by collaborating with them rather than dictating to them. In this more equitable environment, prospective employees and employers seek information about each other. Ideally, they will enter into mutually beneficial relationships characterized by high productivity and the initiative, innovation, and tenure born of true job satisfaction.

Unfortunately, the Army's current officer employment paradigm is not talent driven. Instead, it is industrial (almost feudal) in nature, running counter to best practices. The Army unduly prioritizes "fairness" when making assignments, has a narrowly defined pathway to senior leadership ranks, cannot see the talent it possesses, and suffers from severe principal-agent problems.

The Army must move beyond these industrial era employment practices and adopt information age talent management. However, creating better talent matches requires a significant change in its feudal employment culture. Sound theories, inovative technologies, and controlled market mechanisms can help the

Army match individual officer talents with specific work requirements.

A carefully controlled talent market driven by a state-of-the-art information technology system can help create employment practices equal to our times. It will allow commanders to seek the talent they need, screen job candidates, and interact with both officers and Human Resources Command (HRC) personnel to achieve good matches. In turn, officers will better know what talents are in demand. This can positively shape their developmental decisions, future assignment aspirations, and professional networks.

Most importantly, the Army will benefit on several levels. First, it will finally be able to "see" the talent it possesses and the talent that is actually in demand. As talent gaps are revealed, it will be empowered to allocate officer developmental resources far more efficiently and rapidly. Second, the Army's Officer Corps will work in increasingly networked fashion, building technology-enabled, problem solving relationships. Finally, optimal talent matches will improve talent development, enhance productivity, reduce risk and ensure the Officer Corps has the depth and breadth of talent it needs, both now and in the future.

TOWARDS A U.S. ARMY OFFICER CORPS STRATEGY FOR SUCCESS: EMPLOYING TALENT

INTRODUCTION

In 1911, Frederick Winslow Taylor, a mechanical engineer, published *The Principles of Scientific Management.* His premise was that, in general, workers performed at the slowest rate that goes unpunished, something he (ironically) referred to as "soldiering."[1] To rectify this, Taylor devised a method for improving worker productivity. First, the employer would break skilled labor requirements down into smaller, less skilled labor tasks. Next, the employer would "scientifically" identify the "one best way" to perform these smaller tasks to save time and costs. Workers would then be selected, trained and employed to exacting task standards.

In an era worshipful of science and in the throes of industrialization, scientific management, or "Taylorism" as it came to be called, was a tremendous hit. Bethlehem Steel, Henry Ford, and other manufacturers employed it in their factories, sometimes doubling or tripling output. Even today, this sort of task-oriented work optimization continues in several industries.

The drawbacks of Taylor's program, however, were significant. Chief among them, it failed to recognize that the most efficient way of working for one person might be inefficient for another. It made work repetitive, tedious, and uninteresting. It stifled self-development and smothered employee decisionmaking or innovation. Lastly, it treated people like interchangeable parts, employing just a fraction of their unique talents.

In the Information Age, jobs are becoming more complex, not less so, requiring employees who are agile, inventive, and empathetic. Work is increasingly characterized by high levels of task interdependence, skill specificity, and uncertainty. In addition, today's enormously competitive labor market gives educated professionals the option of seeking new employment whenever a company fails to give them sufficient voice in their work. In short, the industrial era, during which "bosses" unilaterally made employment decisions, is over.

Today, the most successful enterprises unleash the full potential of their workers by collaborating with them rather than dictating to them. In this more equitable environment, prospective employees and employers seek information about each other. Ideally, they will enter into mutually beneficial relationships characterized by high productivity and the initiative, innovation, and tenure born of true job satisfaction.

Employing people optimally is not easy, however. It requires the ability to access the talent in demand, to develop it to meet both current and future demands, and to retain it in an extremely competitive American labor market. If that were not difficult enough, optimal employment engages the critical component of *timing*—getting an employee in position as he approaches the apex of his productive capability *in that position*. By this, we mean that both work requirements and individual talents are always changing—the talent match that may have been optimal 2 or 3 years ago may become less so over time, either because the requirements have changed, the employee has, or both. Organizations therefore cannot become complacent—they must continuously evaluate their talent and their requirements, ensuring that when warranted, people are afforded new opportunities to make optimal work contributions.

Effective talent *employment* is at the core of the Army Officer Human Capital Model—to provide optimally performing officers in all areas (see Figure 1). Getting it right directly supports talent development. It improves job satisfaction, simultaneously increasing talent retention. Moreover, highly productive and satisfied employees are the ultimate recruiting tool, making future talent *accessions* easier.

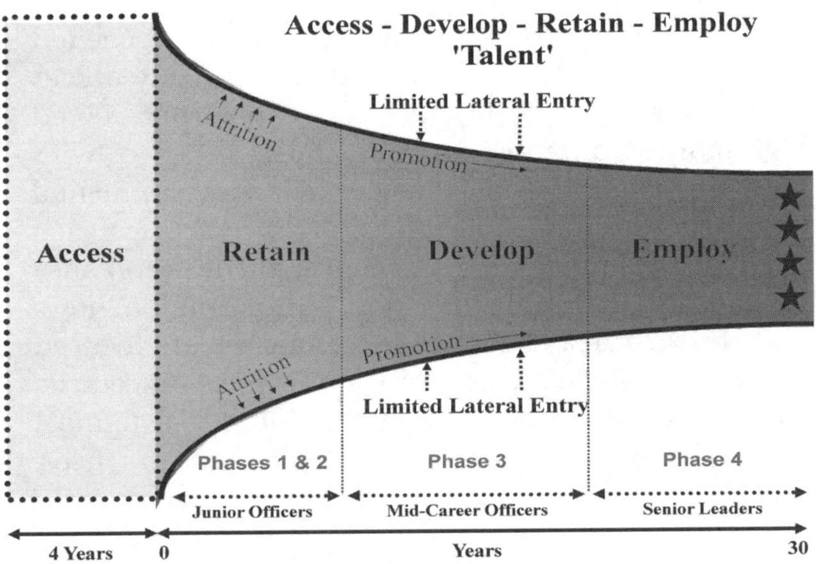

Figure 1. Army Officer Human Capital Model.

In sum, optimal talent employment expands the Army's production possibility frontier—it can do more with existing resources. It also helps ensure that the Officer Corps possesses the depth and breadth of talent needed to meet the twin challenges of a competitive labor market and an increasingly uncertain operating environment.

OFFICER EMPLOYMENT PRACTICES: OUR BIGGEST AREA OF CONCERN

Throughout this monograph series, we have explored several talent management challenges with implications for the future well-being of the Officer Corps. In Volume 3, we discussed the harm caused by low junior officer retention, a challenge now being redressed via the Officer Career Satisfaction Program (OCSP). In Volume 4, we identified accessions practices that not only stunt Army efforts to acquire the officer talent it truly needs, but also rob it of talent needed elsewhere. In Volume 5, we argued that Army officer development practices, which for years have received high marks from most quarters, must keep pace with emerging challenges via changes in its developmental culture, education, and evaluation practices.

All of these talent management challenges are cause for concern, thought, and action. In our opinion, however, the greatest challenge is the one we are focused upon here—the way the Army employs its officers. Its current employment paradigm is industrial (almost feudal) in nature, running counter to best practices. The Army unduly prioritizes "fairness" when making assignments, has a narrowly defined pathway to senior leadership ranks, cannot see the talent it possesses, and suffers from severe principal-agent problems. We will explore each of these challenges in turn.

Fairness.

When an officer hears from Human Resources Command (HRC) about a potential assignment, his or her pulse quickens. It is an understandable response. Assignments dictate where the officer will serve for the next 2-4 years, whom he will serve with, and what he will do. Assignments have an outsized impact upon an officer's future advancement opportunities, as well as upon his or her family and quality of life. Working through it all can be an emotional process.

For the Army, of course, assignments should have no emotional component—they are simply the mechanism through which it derives production from each officer. Yet, in a well-meaning effort to take care of its people, the Army's current officer assignment process focuses much more upon "fairness" than it does upon coolly optimizing officer productivity. Instead of talent considerations, each officer's "dwell" (nondeployed) time, "boots-on-the-ground" ("BOG" or deployed) time, number of deployments, and the number of overseas postings dominate future assignment decisions.

In fact, an HRC branch representative may well begin an officer's assignment interview with this type of a comment: Let's see, you've been in CONUS [in the continental United States] for 3 years—time to get you back in the fight, or: We need to get you an assignment where you can 'take a knee'—you've had two overseas deployments in the last 4 years. However, this way of doing business is problematic, because it short-circuits talent matching, leads to suboptimal productivity, increases risks of mission failure, and demonstrates a skewed notion of fairness.

To be very clear—we support efforts to rest people after challenging or hazardous assignments, to reunite families after extended separations, and to provide equitable deployment exposure. We also wholeheartedly support Army efforts to broaden people (or as we say, extend their talent advantage) by providing them with challenging assignments across a variety of environments. It is necessary to do these things.

However, the practice of weighting deployment exposure more heavily than talent matching when making assignments is terribly shortsighted. It presumes that officers are interchangeable widgets and can therefore be treated identically. As we have argued throughout this monograph series, nothing could be further from the truth. Each officer is a unique individual, possessing a talent set that aligns far better with some assignments than with others.

This is why the Army must recalibrate its notions of fairness. While it must afford *equal* opportunities to all, the *fairest* employment behavior it can engage in is to assign officers where their talents help defeat threats at the lowest cost in American lives and taxpayer dollars. This is true fairness—to the taxpayer, to the Soldiers serving with the officers, and to the Army's joint, interagency, intergovernmental, and multinational partners.

Narrow, Tradition-bound Pathways to Success.

A feudal employment culture can prevent an organization from liberating the talent of its people, particularly in emerging threat or technology areas. During World War I, for example, Brigadier General William "Billy" Mitchell brilliantly commanded all American air combat units in France. At war's end, many expected that General John Pershing would champion Mitchell as the first military director of the Army's Air Service—he was undoubtedly the most talented senior aviation officer in the Army. Instead, Pershing chose Major General Charles Menoher, who had capably commanded the 42nd Infantry Division in France.

Some assume Pershing passed over Mitchell due to his caustic personality.[4] But other factors constrained Pershing's options, chiefly the Army's well-entrenched seniority system. Menoher was an artillery officer and an 1886 West Point graduate, while Mitchell was a "mustang" Signal Corps officer who had received a direct commission 20 years later.[5] In short, General Menoher's source of commission, success as a ground combat branch officer, and far greater seniority all fit the narrow and traditional pathways to senior officership that predominated at the time, even though he had no air service experience.[6] As a result, his assignment was a poor talent match, and Menoher was relieved as Air Service director in 1921. He eventually returned to division and later corps command, where he performed honorably and well.[7]

Pershing's "Mitchell or Menoher" dilemma highlights what can happen when seniority, traditional personnel management techniques, and misplaced notions of fairness supplant talent in the employment process. Such practices can have negative implications at all levels. In this instance, both individual and organizational performance were suboptimized. The Army's Air Service experienced a tumultuous 2 years, during which its director and deputy were continuously at loggerheads. Over the same period, the Army failed to fully benefit from Menoher's talent as a ground forces commander or Mitchell's as an airpower innovator.

Unfortunately, remnants of this century-old employment culture remain in the Army today, restricting its ability to effectively employ officers. As we highlighted in our previous monograph, nearly 80 percent of the Army's senior leader assignments require talent in more than just the operational art.[8] Despite this, the Army's relatively narrow, tradition-bound paths to enterprise leadership heavily transit operational assignments and draw almost exclusively upon "maneuver, fires, and effects" officers (primarily combat arms).

This is appropriate in some instances, of course, but less so in others. As the range of national security threats becomes increasingly asymmetric and nonkinetic, winnowing talent by shunting it down narrow career paths will deny the Army the talent needed to meet those challenges. Success in warfighting, nation-building, disaster relief, and myriad other contingencies requires an organizational breadth of talent that can be sustained only by creating more pathways to enterprise leadership.

An uncertain threat environment also demands a certain *depth* of talent. The Army's existing officer employment practices, however, frustrate the development of depth, particularly for its more senior officers. As officers achieve greater rank and responsibility, their formal development time is increasingly sparse.[9] To redress this, on the job training and experience—*tenure*—becomes critical. This is standard practice in most successful enterprises. Optimally performing employees remain in position long enough to extend their talents and become true innovators. Army culture generally frowns upon tenure, however, characterizing it as "homesteading." This "up and out" employment mindset stifles innovation and hampers the Army's ability to develop deeply talented people.

The Army Cannot See its Talent.

Even if the Army acknowledges that every officer is unique, it will be unable to manage their individual talents until it knows what they are and what talents are needed. Currently, it has little information in this area.

Make no mistake—the Army knows plenty about its officers: their home of record, gender, race, marital status, colleges attended, blood type, and religion. It tracks his health and fitness levels, months deployed, awards, and decorations. It knows many other things as well—the number and type of training courses completed, positions held, dates of promotion, and security clearance levels. All of this information, and more, is found in each officer's record brief (ORB).

Unfortunately, this is simple *accounting* data. To manage officer talent, however, the Army needs *decision support* data, *information that reveals what makes each officer tick.* What does he value? What opportunities does he desire? What incentives will he respond to? What does he know that the Army has not taught him? Where has he been that the Army has not sent him? What does he enjoy? How does he see the future? How does he learn? In other words, what are his (or her) talents?

Ironically, web applications such as *Plaxo, Monster,* or *LinkedIn* often know more about participating officers' talents than the Army does. These networks are flourishing because they motivate people to volunteer vast amounts of professional information via friendly and intuitive user interfaces. As a result, that information is usually current, relevant, and fully searchable, a key advantage over Army personnel information management systems. "Web 2.0" sites are also lightning fast relative to most Army web applications, another advantage. Additionally, they incorporate *inference technology*—the ability to learn about users through continuous interaction and to provide them with increasingly useful and personalized service.

With these tools, civilian employers have gained a real advantage over the Army in the talent wars. Not only can they see each participating officer's talents, but they can attract them to their organizations via detailed job postings. Today's Army officers can use nimble online search tools to find thousands of private sector jobs demanding their talents. This market transparency is in stark contrast to the Army's highly opaque, top-down employment approach, a likely contributor to talent leakage from the Officer Corps.

The "Principal-Agent" Problem.

In addition to knowing which talents it has on hand, the Army must also understand which talents are *in demand* across its organizations. Commanders know which talents they need and officers know which talents they can provide. Unfortunately, neither makes assignments—the Army's HRC does, creating a significant *principal-agent problem*. This arises when two parties do not share the same information and also have differing interests.[10]

In this case, commanders *(the principals)*, are charged with leading their organizations to successful outcomes. They desire "ace" job candidates—officers who can dramatically exceed minimal performance requirements because there is a high correlation between their talents and work requirements. When making assignments, however, HRC's branch managers *(the agents)* have no real mechanism for determining which specific talents commanders are seeking, or how large a supply of it exists in the Officer Corps.

To make matters worse, HRC's interests often lie outside those of commanders. Talented, dedicated, and extremely hard working, HRC's branch managers and assignment officers administer a system seeking a fair distribution of officers, ensuring that each unit shares the same burden of shortages or overages in officer inventory. Under this system, commanders must build their teams with whatever talent HRC assigns to them.

Meanwhile, officers (who are also *principals* in the assignment process) must do their best to perform wherever HRC employs them, whether the job matches their talents or not. Again, we see differing interests. Officers are seeking assignments that liberate their talent and allow them to make an optimal contribution to the Army, while HRC is focused upon a fair distribution of overseas assignments and deployment exposure across the Officer Corps. In a recent survey, however, 44 percent of young officers identified "the job" as their most important consideration when seeking their next assignment. By comparison, only 6 percent of them consider deployment schedules important.[11]

Solving principal-agent problems requires aligning incentives and reducing information asymmetries. Essentially, assignment managers need a way of knowing what talents commanders need and what talents are possessed by the officers they manage. Assignment managers must also be motivated to increase both individual and organizational productivity via information-driven talent matches. Until these issues are resolved, the Army will continue to treat officers as interchangeable parts, suffer low officer retention, endure unnecessarily high developmental costs, and perform suboptimally. Understanding some fundamental theories, however, can help the Army break free of this industrial era employment paradigm and move toward genuine talent management practices.

THEORY-TALENT MATCHING REQUIRES BOTH DATA AND INCENTIVES

The theory of optimal job matching rests upon three key assumptions. First, there is a heterogeneous distribution of both employee talent and employer requirements. Second, there is imperfect information on both sides of any job transaction—neither the employer nor the employee knows whether a good talent match is at hand. And third, there is an incentive mechanism that encourages talent matching for both the employer and the employee.[12]

In our view, these assumptions hold when considering the possibility of a talent-focused Army officer employment system. First, all officers possess varied and unique talent distributions, just as all officer requirements are varied and unique. In fact, the uniqueness of both officers and requirements tends to increase with rank.[13] Second, asymmetric information problems abound—officers have little visibility over the preponderance of jobs for which they might be a great talent match, and the Army knows very little about the talent of each officer. Finally, it is in the best interest of both the Army and individual officers to match talents against requirements. The organization increases its productivity without increased costs, and the officer experiences enhanced productivity and job satisfaction without compromising his or her career.

We can conceptualize the methods for achieving talent matches as lying along a continuum, from "command directed" to "market driven" in nature. In our daily lives, we are surrounded by evidence that the operation of markets (with appropriate safeguards in place) engender far more efficient and productive outcomes than command directed processes do.

Recent world history reinforces the point. Compare the U.S. and Soviet economies, for example. In 1945, these two global superpowers both possessed significant quantities of heavy industry, natural resources, labor, etc. By 1990, however, the Soviet Union's state-planned economy was barely one-third the size of the American economy. In fact, the gap between the two had been growing *wider* for years, despite Soviet predictions that their industrial production and per capita income would eclipse that of the United States by 1980.[14]

Like the old Soviet economy, a rigid, centrally managed approach to employing officers is woefully inefficient and unequal to the needs of today's volunteer force. It requires the Army to know *exactly* what its future talent requirements will be—an impossible task. Nearly as impossible, it tells people what they will do and expects them to perform optimally in *any* assignment they receive. This approach puts a premium on having adaptable (interchangeable) officers.

At the other end of the continuum is a regulated, market-driven employment approach that would create incentives for officers (the labor supply) to volunteer talent information and for commanders (the labor demand) to identify talent requirements. In this way, the Army could wean itself from reliance upon error-prone requirements forecasts. Instead, it could become a truly agile enterprise, better employing people within their unique talent sets. The Army's Officer Corps might then achieve genuine breadth and depth of capability without requiring every officer to master everything (the pentathlete paradigm).

To illustrate the way in which market forces can help organizations meet unforeseen and rapidly emerging talent requirements, consider Figure 2, which compares undergraduate Middle Eastern studies by West Point cadets with graduate-level Middle Eastern studies by Army officers.

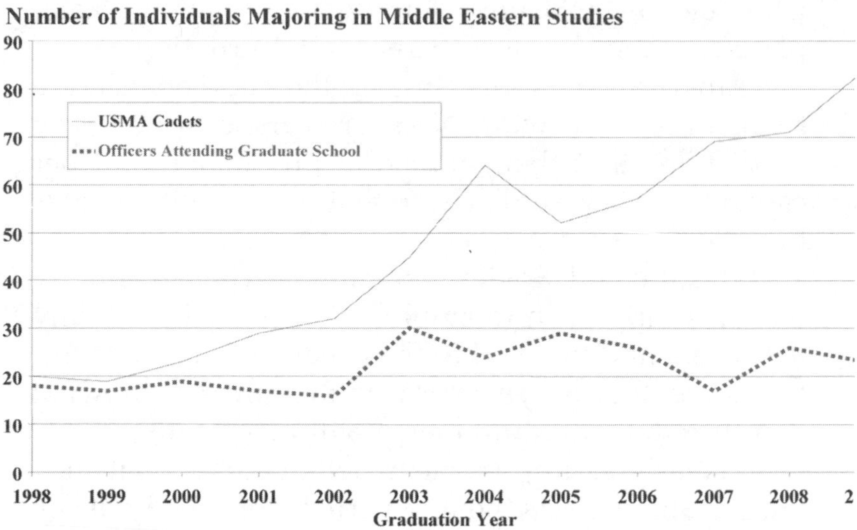

Figure 2. Individuals in a Free Market Respond More Rapidly to Changing Demand than Command-Directed Enterprises Can.

Just as at any American university, West Point cadets can choose their programs of study. The solid line shows how quickly they responded to the events of September 11, 2001 (9/11). Almost immediately, the number of cadets choosing Middle Eastern studies increased dramatically. An incentive is in play—young men and women embarking on an Army career want to bring relevant talents to their profession. The Army (via West Point) affords these young people with the opportunity to extend their talents. In return, it gains much needed capability from people with both the talent and the desire to provide it. Both parties to this exchange benefit rapidly and tremendously.

If every cadet wanted to major in Middle Eastern studies, West Point would have to regulate this market because the Army still requires engineers, economists, historians, and experts in other regional studies. To date, however, there has been no need for intervention in the selection of majors by cadets—the market clears optimally.

In stark contrast to the example cited above, graduate level programs of study for Army officers are centrally controlled and allocated. The dotted line in Figure 2 tells the story. During almost a decade of persistent conflict in the Middle East, the Army did not increase the number of officers enrolled in graduate-level Middle Eastern studies. Perhaps this was due to internal debate over the wisdom of doing so: Which program study areas do we curtail if we allow more officers to study the Middle East? Regardless, the Army did not react, and an opportunity to increase its cultural fluency in a critical area was lost.[15]

Top-down, centrally managed human capital practices may have been sufficient during the relative equilibrium of the Cold War era, with its industrial economies, conscript armies, and clear adversaries. They are unequal, however, to the needs of a volunteer force facing the twin challenges of a competitive labor market and an increasingly complex global operating environment. Moreover, they are unnecessary.

Information age tools make it possible to capture a great deal of information regarding individual talents and unique work requirements, while market mechanisms can help the Army use that information with telling effect. Instead of trying to forecast, for example, how many electrical engineers the Officer Corps needs, the Army will know based upon the actual demand for that talent set.[16]

In addition, as Army talent demands become clear, officers will be better able to develop the skills to meet them. In cases where jobs require particular depth or specialization, the Army may also consider extending tenure to officers, both to increase their on-the-job development and to reap the highest rate of return from extremely productive individuals with rare talents.

Market mechanisms incentivize employees and employers to provide granular data on their respective talents and requirements. This is critical to creating optimal job matches. The more granular the information, the greater the advantage one potential employee has over another for a particular job. Accuracy is encouraged as well—careless mistakes or deliberate falsification of information can lead to poor job matches that effectively end an officer's career.

This level of detailed information can introduce an entirely new component to officer evaluations. Currently, all officers, regardless of rank, position, branch, location, tenure, span of responsibilities, etc., are evaluated against identical performance measures via the Officer Evaluation Report (OER). However, future evaluations will be able to go much further.

Using detailed information about an officer's talent and the job's specific requirements, commanders and personnel managers will assess not just performance but *the strength of the talent match*. Was the job a good fit? If not, why not? How was the officer selected for this position? What information was used to make this assignment? What credentials are needed to succeed at this job in the future?

Today, when an officer fails to perform optimally, the Army holds the officer responsible, and the implications for his or her career can be serious. In the future, however, the assessment might read, "We put him in the wrong job, now let's get it right."

TOWARDS A TALENT MANAGEMENT APPROACH: GREEN PAGES

To test the theories described above, an innovative new web application is currently being piloted on a small scale among Engineer officers. Called simply "Green Pages," it is more than just a talent-matching or employment tool.[17] Green Pages proceeds from an understanding of how markets work, why they fail, and how they can be regulated. It also draws upon behavioral economic theory—how people behave in a marketplace and which incentives will move them to action.

Currently, there is no market for officer talent in the Army—no way for organizational strength managers and individual officers to make efficient talent transactions. This represents a *market failure*—an inefficient use of resources when better results are possible. In other words, assignment transactions still occur, but there is a significant misalignment of talent supply and demand, making the Officer Corps less productive than it can be. Green Pages can rectify this, providing the Army with its first market-driven officer talent management system.

Operating Concept.

Figure 3 graphically depicts the Green Pages operating concept, simple in design but potentially quite powerful in implementation:

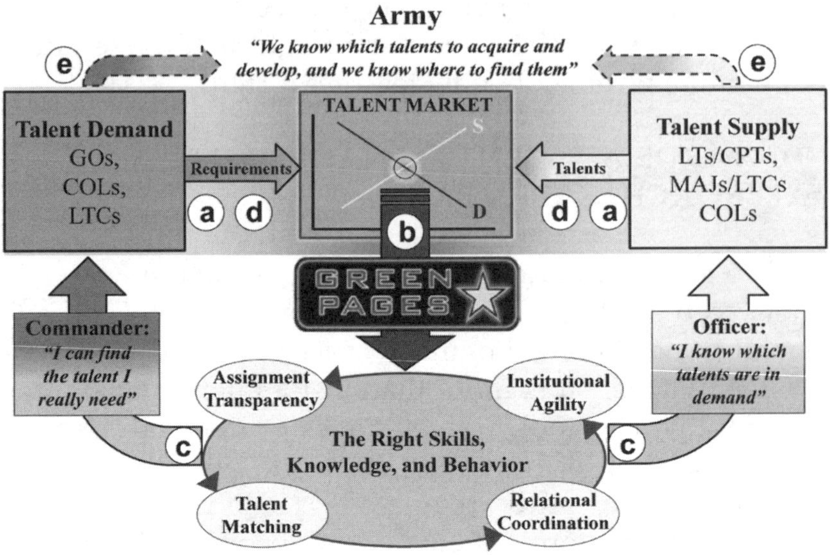

Figure 3. Green Pages Reveals Both the Talents the Army Possesses and the Talents It Demands.

Each person's collective life experiences represent tremendous capital in the Army talent market. When officers participate actively in Green Pages (Figure 3, point a), they will create detailed profiles summarizing all of their expertise, experiences, and accomplishments. More than just a listing of Army training and skill identifiers, these include talents gained in college, through leisure pursuits and hobbies, in their communities, in the civilian job market, and even from relationships with friends and family.

The Engineer pilot currently underway provides excellent examples of the new officer information that Green Pages is revealing, everything from what officers can do to what they *hope* to do.[18] Examples of actual information already entered into the system include:

- A captain who wishes to obtain his professional engineering and Leadership in Energy and Environmental Design (LEED) certifications, and plans on taking the Fundamentals of Engineering Exam (FEE) this summer while pursuing his masters degree in environmental engineering.
- A lieutenant who interned throughout college with an engineering firm building light rail systems in the Southwest. As a civilian, he also owned and ran a "green" business.
- A lieutenant who has extensive prior experience as a project design and construction manager in the Baltimore, MD, and Washington, DC, metropolitan areas, as well as abroad.

These officers are sharing talents and goals in a professional setting, information which has tremendous assignment utility and is not available to the Army elsewhere. As you can imagine, a flood of other officer data, all searchable, will continue to enter Green Pages, such as professional journal articles written, heritage languages learned in childhood, productive hobbies, publications, contingency experience, etc.

Simultaneously, commanders and strength managers at organizations across the Army will post robust job profiles, detailing not just *required* talents, but *desired* talents. Just as individuals are unique, so are work requirements. Even seemingly identical jobs can differ based upon a variety of factors. These include leadership styles, talent gaps, unit mission, other contingencies, geography, equipment, operating theater, rules of engagement, etc.

The Green Pages Engineer pilot is making this abundantly clear. For example, a captain's position at Camp Zama, Japan, previously identified only by title, rank, and branch, now includes desired civilian educational levels and academic disciplines. It describes the work to be performed and the mission of the organization. It identifies professional certifications that will help an officer excel in the assignment. It provides a point of contact and a website where a job candidate can learn more. Perhaps most importantly, officers can consider their suitability for this job, whereas previously they may not have known it existed.

Once detailed personal profiles and job profiles are entered into Green Pages, they will form the basis for a talent marketplace (Figure 3, point b). As officers and organizations search against one another, the bulk of the talent market will "clear" optimally. In other words, less intervention will be required by actors outside of the talent transaction. HRC will still be the assignments arbiter but will hopefully find itself more in the role of advocating for and approving talent matches rather than balancing officer shortages and deployment exposure.

Workforce talent matching will make the Army more productive as officer talents are liberated by assignments that fit better than previously possible. This will enhance relational coordination through teamwork and collaboration via information networks and face-to face interaction. It will increase job satisfaction, which has direct implications for retention. It will also provide greater organizational agility as units gain the breadth and depth of talent required to succeed in an increasingly complex operating environment (Figure 3, point c). Green Pages will also allow officers to look beyond their next assignment, to know what talents are in demand, align this information with their personal career preferences, and make the developmental choices that will posture them for the assignments they desire in the future.

The talent market created by Green Pages will be dynamic, both iterative and continuous, as new talents and new requirements are continuously fed into the marketplace (Figure 3, point d). As granular information on the Army's talent supply and demand emerges, the Army can abandon static forecasting. Instead, it will see in real time where its talent surpluses and shortfalls are and can rapidly adjust its accessions, development, retention and employment practices (Figure 3, point e).

Core Capabilities.

Green Pages is a "Web 2.0" application, and functionality is benchmarked from the best commercial professional networking applications. It moves beyond those applications, however, which rely almost exclusively upon user input ("how I see and represent myself") rather than official records ("how others see me"). Green Pages combines both user entry information *and* official file information into a comprehensive and searchable profile.

Green Pages also allows users to: manage the information that is publicly available about them as professionals; search against *every* officer position in the Army inventory; contact organization personnel strength managers for more information; be found by Army organizations conducting talent searches; collaborate with fellow experts from across the Army to gather data, share files and solve problems; gain new insights from discussions with like-minded professionals in private group settings; build professional networks that can help them land the jobs they want in the places the Army wants them; and, at the organization level, post and distribute job listings to find and attract the best talent available.

Importantly, Green Pages is a relational database tool, currently fed by several Army data sources. Over time, it can easily draw upon additional data sources to expand its searchable talent information, becoming an increasingly more powerful tool in the process.

Changing Culture and Practice.

While we have described the more immediate benefits of Green Pages, those benefits will likely deepen as the Army's employment paradigm gradually shifts from feudal to collaborative, from exclusively command-directed to increasingly market-driven. Over time, Green Pages can usher in beneficial changes in the Army's work culture and practices.

By giving commanders greater voice in who is assigned to their organizations, for example, a regulated talent market supported by Green Pages can help the Army truly make Soldiers its centerpiece. Consider that today's commanders do not bear the cost of labor because it is "loaned" to them by an outside agent (HRC). They take what they get and make do. As a result, in today's Army culture, commanders are held more accountable for the operational readiness of their pacing items than they are for the long-term career viability of their officers.

If a battalion commander averaged a 70 percent operational rate for his tank fleet, he would leave command with his career in tatters. However, if 70 percent of his junior officers left the Army at the end of their active duty service obligation (ADSO), there would be no career repercussions for him at all. Why should there be? It is entirely possible that these officers arrived to his unit fully intending to leave the service, or perhaps were terrible matches for his organization. If the bulk of these junior officers, however, were assigned to a unit because of their desires and the commander's wishes, the equation (and the Army's culture) would change. A moral contract is created, and the commander is now responsible for developing and employing young people that are serving with him at his request. He has personally built the team, and his investment in its success on a human as well as operational level rises dramatically.

Green Pages may also change work practices by engendering far greater relational coordination—frequent, timely, accurate, problem solving communication, connecting Soldiers around the world and across time zones and operating theaters. Green Pages provides secure For Official Use Only (FOUO) communications tools: an internal email client, a professional "Answers" module, and the ability to join "Groups" and build a trusted network of associates.

Imagine serving as an engineer construction officer in Mosul, Iraq where you must drill several wells. You have PDC bits, but due to unanticipated soil conditions you need a steady supply of roller cone bits. Several local contractors sell them but they are of poor quality and wear rapidly. Via Green Pages, however, you are able to quickly locate an officer at Fort Lewis who faced a similar challenge two years ago. He informs you of a great local supplier, one you were unaware of. You make contact, secure high quality roller cone bits and triple your drilling speed, all because a simple web application provided you with a rapidly searchable knowledge network to fall back upon. You also become acquainted with an officer you never would have known—his assistance is just the beginning of years of professional collaboration between the two of you. Networked problem solving brings remarkable organizational agility to the Army.

When natural disasters strike, such as the recent earthquake in Haiti, Green Pages can help the Army assemble the most talented response team possible. A commander can immediately search for people by cultural fluency, law enforcement, engineering, or any other work requirements. He or she can search not just official records, but officer-provided information revealing relevant talents gained via leisure travel, a religious mission, a Peace Corps stint, a Habitat for Humanity project, advanced civil study, training with industry, civilian employment, etc.

Perhaps more importantly, Green Pages may eventually span branch and component boundaries that can be barriers to talent employment. Imagine that the Army is responding to another Katrina-like hurricane in the Gulf of Mexico. Unlike 2005, this time Green Pages immediately identifies all engineers with levee building and reconstruction experience. The Army promptly dispatches these officers to the New Orleans Corps of Engineers district. However, a talent gap emerges—there are more officers needed than available. Another Green Pages search takes place, this time focused upon specific engineering talents and experience, rather than just Active Component "engineer branch" officers. The search reveals several certified engineers with the required geotechnical experience and credentials, to include one in the Mississippi National Guard (an Infantry officer), another in the Army Reserve who lives in Maryland (an MP officer), and several others. All are mobilized and deployed to the crisis zone.

Potentially, Green Pages can reach all the way back into new accessions to ensure the Army fully leverages the talents of its junior officers and places them upon the most productive and rewarding career paths possible. For example, via Green Pages the Army could make officer branching decisions based upon far more information than is available today (and on both sides of the market—talent and requirements). Just as college graduates prepare resumes and interview with civilian employers, prospective officers could engage in a similar process with the basic branch they feel best matches their talents.

CONCLUSIONS

Talent *employment* is at the core of the Army Officer Human Capital Model. The Army's current employment paradigm, however, is unequal to the needs of a professional volunteer Army facing the twin challenges of a competitive labor market and an increasingly complex global operating environment. It unduly prioritizes "fairness" when making assignments, has a narrowly defined pathway to senior leadership ranks, cannot see the talent it possesses, and suffers from severe principal-agent problems.

The Army must move beyond industrial era employment practices and adopt information age talent management. However, creating better talent matches requires a significant change in its feudal employment culture. Sound theories, information age tools, and controlled market mechanisms can help the Army match individual officer talents with specific work requirements.

A carefully controlled talent market driven by Green Pages is a win-win proposition. *Commanders* win because they can seek the talent they need, screen job candidates, and interact with both officers and HRC personnel to achieve good matches. *Officers* win because they will better know what talents are in demand. This can positively shape their developmental decisions, future assignment aspirations, and professional networks.

The *Army* wins as well, and on several levels. First, it can finally see the talent it possesses and the talent that is actually in demand. As talent gaps are revealed, it can allocate officer developmental resources far more efficiently and rapidly. Second, the Army's Officer Corps will work in increasingly networked fashion via Green Pages, building technology-enabled problem solving relationships. Lastly, optimal talent matches will improve talent development, enhance productivity, reduce risk, and ensure the Officer Corps has the depth and breadth of talent it needs, both now and in the future.

ENDNOTES

1. Frederick Winslow Taylor, *The Principles of Scientific Management*, New York: Harper and Brothers, 1911, p. 19.

2. See "Frederick Taylor and Scientific Management," Net MBA Business Knowledge Center, available from *www.netmba.com/mgmt/scientific/*.

3. See Casey Wardynski, David S. Lyle, and Michael J. Colorusso, *Talent: Implications for a U.S. Army Officer Corps Strategy*, Vol. 2 of this monograph series, Carlisle, PA: Strategic Studies Institute, U.S. Army War College, November 2009, p. 13, available from *www.officer-strategy.strategicstudiesinstitute.army.mil/sessions.php?catid=2*.

4. While beloved by his subordinates, Mitchell had little tolerance for superiors who did not share his airpower vision. He could be thorny to the point of insubordination (a factor that would result in his court-martial several years later). He was also flamboyant, viewed by many senior officers as too free-speaking and self-promoting.

5. A "mustang" officer is one commissioned directly from the enlisted ranks.

6. Challenging the Army's heavily entrenched seniority system was not something Pershing was likely to take on, particularly as his own career had benefitted from such a challenge, creating deep resentments within the Army that endured across his career. In 1903, Pershing was a 43-year-old captain who had caught the attention of President Theodore Roosevelt as a vigorous and enterprising officer. Roosevelt petitioned the Army's General Staff to advance Pershing to field grade rank, but the Army refused to upset its seniority system. Roosevelt overcame this by exercising presidential prerogative, nominating Pershing as a brigadier general. Congress approved the nomination in 1906. This catapulted Pershing over more than 800 senior officers, shocking the Army establishment. Many officers considered the promotion the result of timely political patronage rather than merit. In addition to Roosevelt's support, Senator Francis Warren, Chair of the Senate's Military Appropriations Committee, had become Pershing's father-in-law in 1905. See Matthew M. Oyos, "Theodore Roosevelt, Congress and the Military: US Civil-Military Relations in the Early Twentieth Century," *Presidential Studies Quarterly*, Vol. 30, No. 2, June 2000, pp. 312-330, available from *www.jstor.org/pss/27552096.*

7. To his credit, Major General Menoher fully recognized that he was unprepared to serve as Air Service chief, which is why he had requested Brigadier General Mitchell's appointment as his deputy. See *en.wikipedia.org/wiki/Charles_T._Menoher.*

8. See Casey Wardynski, David S. Lyle, and Michael J. Colorusso, *Towards a U.S. Army Officer Corps Strategy for Success: Developing Talent,* Vol. 5 in this monograph series, Carlisle, PA: Strategic Studies Institute, U.S. Army War College, March 2010, p. 13.

9. *Ibid.*, pp. 8-9.

10. See *en.wikipedia.org/wiki/Principal-agent_problem.*

11. Source: "Army Green Pages Pilot Program, Survey of BOLC B and Captains Career Course Students", Fort Leonard Wood, MO, March 2010. In addition, 97 percent of respondents believe that the creation of detailed officer job profiles within units is "very important" or "somewhat important."

12. Boyan Jovanovic, "Job Matching and the Theory of Turnover," *Journal of Political Economy,* Vol 87, 1979, p. 974.

13. In our view, the depth and breadth of individual talent grows during a career. Correspondingly, positions of greater responsibility generally require increasingly specific talents. This means that the utility of talent matching rises for mid to senior-ranking officers, although it is beneficial at all levels.

14. All Business, "How Big was the Soviet GDP?" available from *www.allbusiness.com/government/630097-1.html.* Also see Gordon M. Hahn, "An Autopsy of the Soviet Economy," Stanford, CA: Hoover Institution, Stanford University, available from *www.hoover.org/publications/digest/3540681.html.*

15. The Army Education Requirements System (AERS) is governed by Army Regulation 621-108. This regulation calls for an annual review and validation of all positions requiring an advanced degree, with approval granted by HRC and the Army G1.

16. Hedonic demand theory suggests that the market will reveal information about the true object of demand through the supply and demand mechanism. It disaggregates talent into its constituent characteristics in an effort to determine the contributory value of each characteristic. In other words, it is the demand for the *characteristic,* not the demand for the individual *possessing* the characteristic, which reveals the valuable information. See *www.probertencyclopaedia.com/cgi-bin/res.pl?keyword=Demand+Theory&offset=0.*

17. Green Pages was chosen as the name for this application because it evokes the Army while also harkening back to yellow and white pages, sources of information on organizations and individuals.

18. Green Pages is being piloted with the Engineers because the Chief of Engineers requested it. Over the last decade, the branch has been sorely tested not just by war, but by simultaneous crises and humanitarian relief missions resulting from natural disasters in the U.S., Asia, the Caribbean, etc. As demand for engineer officer talent has surged, the Engineers realized that the current way of managing officers has not allowed them to respond as effectively as possible.